Bearded Dragon Food

Nutritional Needs And Feeding Guide For Bearded Dragons

Raymond Jack Tyler

Table of Contents

CHAPTER ONE 4
- INTRODUCTION 4
- NATURAL DIET OF BEARDED DRAGONS 6
- COMMERCIAL BEARDED DRAGON FOODS................ 9

CHAPTER TWO...13
- FRESH VEGETABLES AND FRUITS............................13
- LIVE FOOD CHOICES..16
- SUPPLEMENTS AND VITAMINS 18

CHAPTER THREE 21
- FEEDING SCHEDULE FOR BEARDED DRAGONS 21

CHAPTER FOUR 29
- TYPES OF BEARDED DRAGON DIETS 29

CHAPTER FIVE...38
- SIGNS OF NUTRITIONAL IMBALANCE 38
- FEEDING YOUNG VERSUS GROWN-UP BEARDED DRAGONS.44

CHAPTER SIX...50
- HYDRATION AND WATER NEEDS.............................50
- OCCASIONAL DIET CHANGES 53
- PICKING AND RAISING LIVE FEEDERS 56

CHAPTER SEVEN 61
- IMPORTANCE OF GUT LOADING AND DUSTING ...61
- FEEDING BEHAVIOR AND FRQUENCY 64

CHAPTER EIGHT..69
- COMMON NUTRITIONAL DEFICIENCIES AND SOLUTIONS.....69

FEEDING CHALLENGES AND SOLUTIONS 73

CHAPTER NINE 78

SIGNS OF OVERFEEDING AND UNDERFEEDING 78

THE END 83

CHAPTER ONE

INTRODUCTION

Bearded dragons are interesting reptiles that have acquired prevalence as pets because of their quiet nature and generally simple care necessities. Proper nutrition is important for the health and well-being of these animals. In the wild, bearded dragons are omnivores, meaning they eat a variety of bugs, vegetables, and fruits. As pets, reproducing this natural diet is fundamental to guaranteeing they get every one of the vital nutrients. Be that as it may, numerous proprietors likewise depend on commercial bearded dragon foods to enhance or supplant portions of their diet.

Bearded dragons require a fluctuating and balanced diet to flourish in captivity. Imitating their natural diet of bugs, vegetables, and infrequent fruits is critical for their health and well-being. Commercial bearded dragon foods can be a helpful and compelling method for guaranteeing they get every one of the vital nutrients, but they shouldn't supplant new foods totally. By combining new, natural foods with high-quality commercial choices and suitable enhancements, you can furnish your bearded dragon with a nutritionally complete and fulfilling diet. Remember, consistently screen your pet's health and change their diet on a case-by-case basis to guarantee they stay cheerful and healthy all through their life.

NATURAL DIET OF BEARDED DRAGONS

Right at home in Australia, bearded dragons have a different diet that consists of both insects and plant matter. Here is a breakdown of what wild bearded dragons commonly eat:

1. Insects:

a. Bearded dragons are insectivores when they are young, and bugs play a critical role in their diet all through their lives.

b. Common bugs in their diet incorporate crickets, mealworms, waxworms, and cockroaches.

c.   These bugs provide fundamental proteins and fats vital for development and, generally speaking, health.

2.   Vegetables and Greens:

a.   As bearded dragons get older, their diet shifts towards being more herbivorous.

b.   They consume a variety of vegetables and greens, for example, collard greens, mustard greens, kale, and dandelion greens.

c.   These salad greens are rich in vitamins and minerals, including calcium, which is essential for bone health.

3.   Fruits:

a. Fruits are a more uncommon but still important piece of a bearded dragon's diet.

b. They can eat fruits like berries, melons, and apples as intermittent treats.

c. Fruits provide natural sugars and extra vitamins.

4. Flowers and Plants:

a. Bearded dragons likewise consume a variety of flowers and plants in the wild.

b. Edible flowers like hibiscus and dandelions can be proposed to them as a component of a fluctuating diet.

To duplicate this natural diet in captivity, proprietors ought to plan to give a balanced blend of bugs, vegetables, and periodic fruits to guarantee their bearded dragons get every one of the vital nutrients.

COMMERCIAL BEARDED DRAGON FOODS

While it's important to repeat the natural diet of bearded dragons, numerous proprietors likewise utilize commercial bearded dragon foods as a helpful method for guaranteeing their pets are getting a balanced diet. These commercial foods come in different structures, including pellets, canned foods, and freeze-dried choices. This is the thing you want to be familiar with commercial bearded dragon foods:

1. Nutritional Balance:

a. High-quality commercial foods are figured out to give a balanced blend of proteins, fats, vitamins, and minerals.

b. Look for foods that run down bugs and vegetables as essential fixings to guarantee they intently emulate a bearded dragon's natural diet.

2. Convenience:

a. Commercial foods can be a helpful choice for busy proprietors or as an enhancement to a new diet.

b.   They have a longer usability time span and can be simpler to store than new foods.

3.   Variety:

a.   Some commercial foods are intended to offer a variety of surfaces and flavors, which can assist with forestalling dietary boredom and empower natural searching behaviors.

4.   Supplements:

a.   While commercial foods mean to be nutritionally finished, it's as yet fundamental to give extra calcium and vitamin enhancements to guarantee your bearded dragon's health.

b.   Dusting bugs with calcium powder and, at times, adding a multivitamin supplement to their diet can assist with filling any nutritional holes.

CHAPTER TWO

FRESH VEGETABLES AND FRUITS

A significant part of a bearded dragon's diet ought to consist of fresh vegetables and fruits. These provide fundamental vitamins, minerals, and fiber that are crucial for their general health. Here are a few prescribed vegetables and fruits to remember for your bearded dragon's diet:

1.   Leafy Greens:

a.   Collard greens

b.   Mustard greens

c.   Turnip greens

d,   Dandelion greens

e. Kale

f. Romaine lettuce (with some restraint)

Mixed greens ought to make up a huge part of the vegetable part of your bearded dragon's diet. They are rich in calcium and other fundamental nutrients. Turn the greens regularly to give them a variety of nutrients and to forestall finicky eating.

2. Other Vegetables:

a. Bell peppers (all colors)

b. Carrots (periodically)

b. Squash (butternut, oak seed)

c. Zucchini

d.   Cucumber

These vegetables can be presented close by salad greens to give a balanced diet. Remember to slash or mesh the vegetables into manageable sizes for your bearded dragon.

3.   Fruits:

a.   Berries (strawberries, blueberries, raspberries)

b.   Apples (eliminate seeds)

c.   Melons (melon, honeydew)

d.   Mango

e.   Papaya

Fruits ought to be offered sparingly, as they are high in sugar. Limit fruit consumption to a couple of times each week and guarantee that the pieces are properly estimated to forestall stifling.

LIVE FOOD CHOICES

Notwithstanding fresh vegetables and fruits, bearded dragons require live prey to meet their protein and fat necessities. Here are some suitable live food choices for your bearded dragon:

1. Insects:

a. Crickets

b. Dubia insects

c. Mealworms (with some restraint)

d. Superworms (grown-ups as it were)

e. Black fighter fly hatchlings

While offering bugs, guarantee they are properly estimated for your bearded dragon. Abstain from feeding wild-got bugs, as they might transmit parasites or pesticides.

2. Other Prey:

a. Phoenix worms

b. Silkworms

c. Hornworms

These prey things can be presented as incidental treats or to change up your bearded dragon's diet.

It is important to destroy the burden and residue of live prey with calcium and vitamin supplements before feeding them to your bearded dragon. Stomach stacking includes feeding the bugs a nutritious diet before offering them to your pet, while cleaning includes softly covering the bugs with calcium or vitamin powder.

SUPPLEMENTS AND VITAMINS

Supplements and vitamins are fundamental for guaranteeing that your bearded dragon gets every one of the essential nutrients for ideal health. Here are the critical enhancements and vitamins to remember for your bearded dragon's diet:

1. Calcium:

a.   Calcium is indispensable for bone health and muscle capability.

b.   Offer calcium powder (without vitamin D3) at each meal.

c.   Use calcium with vitamin D3 two times every month to assist with calcium absorption.

2.   Multivitamins:

a.   A multivitamin supplement can assist in guaranteeing your bearded dragon gets every one of the fundamental vitamins and minerals.

b.   Offer a multivitamin supplement one time each week, rotating calcium with vitamin D3.

3. Probiotics:

a. Probiotics can assist with maintaining a healthy digestive system and preventing digestive issues.

b. Offer probiotics one time per week, either in powder form or through natural sources like unsweetened yogurt.

4. Omega Fatty Acids:

a. Omega fatty acids are beneficial for skin and, overall health.

b. Offer as an enhancement or through foods like fish and flaxseed.

CHAPTER THREE
FEEDING SCHEDULE FOR BEARDED DRAGONS

Establishing a proper feeding plan for your bearded dragon is urgent for its health and well-being. A balanced diet is fundamental to guaranteeing that your pet gets every one of the essential nutrients it needs to flourish. Here is a straightforward and compelling feeding timetable to assist you with keeping your bearded dragon healthy and cheerful.

Day to day Feeding Schedule

Morning:

1. 8:00 AM: Offer fresh vegetables and fruits

2. 9:00 AM: Give live insects (crickets, mealworms, or insects)

Afternoon:

1. 1:00 PM: Eliminate any uneaten vegetables and fruits

2. 2:00 PM: Offer all the more live insects

Evening:

1. 6:00 PM: Last feeding of the day, offer a blend of vegetables and insects

2. 7:00 PM: Eliminate any uneaten food

Week after week Schedule

Monday:

1. Vegetables: Mustard vegetables, zucchini, and collard greens

2. Insects: Crickets

Wednesday:

1. Vegetables: Kale, bell peppers, and carrots

2. Insects: Mealworms

Friday:

1. Vegetables: Dandelion greens, zucchini, and yam

2. Insects: Roaches

Sunday:

1. Vegetables: Turnip greens, green beans, and butternut squash

2. Insects: Crickets

Things to Remember:

1. Continuously give fresh, clean water in a shallow dish.

2. Screen your bearded dragon's weight and change the feeding amounts as needed.

3. Try not to feed an excessive number of fatty insects, like waxworms, as they can prompt obesity.

Common Food Errors for Bearded Dragons

Feeding your bearded dragon some unacceptable foods or in improper amounts can prompt serious health issues. Stay away from these common food slip-ups to guarantee your pet's well-being.

1. Feeding an excessive number of Insects

While bugs are an important piece of a bearded dragon's diet, they shouldn't make up most of their meals. A diet that is too high in protein can prompt liver and kidney problems. Adhere to a diet where bugs make up about 20–30% of their all-out food consumption.

2. Offering Incorrect Vegetables and Fruits

Certain foods and vegetables are not good for bearded dragons. Try not to feed your pet foods that are high in oxalates, similar to spinach, as they can bind calcium and lead to calcium insufficiency. Stick to mixed greens like collard greens, mustard greens, and dandelion greens, as well as other safe vegetables like squash, bell peppers, and carrots.

3. Not Giving Sufficient Calcium and Vitamin Enhancements

Bearded dragons require calcium and vitamin enhancements to maintain strong bones and overall health. Tidying their food with calcium powder and offering a multivitamin supplement two or three times each week is fundamental to preventing metabolic bone disease.

4.   Feeding Processed Foods

Try not to feed your bearded dragon processed foods, including commercial bearded dragon pellets. These foods frequently lack the necessary nutrients and can prompt obesity and other health issues. Adhere to a diet of fresh vegetables, fruits, and live insects to guarantee a balanced and nutritious diet.

5.   Overfeeding

Overfeeding can prompt obesity and other health issues in bearded dragons. Follow the feeding schedule and screen your pet's load to guarantee that you are not overfeeding. A decent guideline is to feed your bearded dragon as much they can eat in about 10–15 minutes, and eliminate any uneaten food to forestall overfeeding.

6. Disregarding Hydration Needs

Bearded dragons require proper hydration to remain healthy. Continuously give fresh, clean water in a shallow dish and fog their enclosure regularly to keep up with the humidity levels. Absence of proper hydration can prompt parchedness and other health issues.

CHAPTER FOUR

TYPES OF BEARDED DRAGON DIETS

Bearded dragons are captivating reptiles that have acquired notoriety as pets because of their unique appearance and moderately tame nature. Proper nutrition is significant for their health and well-being. Understanding the various kinds of diets available for bearded dragons is fundamental to furnishing them with a balanced and nutritious meal plan.

1. Insect Based Diet

The insect based diet is a staple for young bearded dragons and is fundamental for their development and improvement. In the wild, young bearded dragons are fundamentally insectivores, meaning they predominantly eat insects. As they mature, their diet becomes more fluctuating, but insects remain an important piece of their nutrition.

Types of Insects:

a. Crickets: Crickets are one of the most commonly fed insects to bearded dragons. They are rich in protein and can be gut-loaded with nutritious foods to improve their nutritional incentive for the dragon.

b. Dubia Insects: Dubia cockroaches are one more superb decision for bearded dragons. They are high in protein and somewhat simple to process.

c. Mealworms: While mealworms can be fed to bearded dragons, they ought to be presented with some restraint because of their high-fat substance. They can be a decent intermittent treat, but they ought not be the principal staple.

d. Superworms: Superworms are bigger than mealworms and can be fed to grown-up bearded dragons. Similarly, as with mealworms, they ought to be fed sparingly because of their high fat content.

Feeding Guidelines:

a.   Size of Prey: The size of the insects ought to be suitable for the size of the bearded dragon. Young dragons can be fed more modest insects, while grown-up dragons can consume bigger insects.

b.   Frequency: Young bearded dragons ought to be fed insects day to day, while grown-up dragons can be fed insects 2-3 times each week.

c.   Goat-Loading: It is fundamental to gut-load the insects with nutritious foods before feeding them to the bearded dragon. This guarantees that the insects are loaded with fundamental nutrients.

2. Plant-Based Diet

As bearded dragons mature, their diet becomes more plant-based. In the wild, grown-up bearded dragons are omnivores, meaning they eat both plants and insects. A plant-based diet is fundamental for furnishing grown-up bearded dragons with important vitamins and minerals.

Types of Vegetables:

a. Mixed Greens: Kale, collard greens, mustard greens, and dandelion greens are incredible decisions for bearded dragons. They are rich in calcium and other fundamental nutrients.

b. Squash and Pumpkin: Butternut squash, oak seed squash, and pumpkin can be fed to bearded dragons. They are high in vitamins and minerals and are a decent source of hydration.

c. Bell Peppers: Bell peppers are rich in vitamin C and can be fed to bearded dragons as a feature of their vegetable admission.

d. Carrots: Carrots are a decent wellspring of beta-carotene and can be fed to bearded dragons with some restraint.

Feeding Guidelines:

a. Variety: It is vital to offer a variety of vegetables to guarantee that the bearded dragon gets a balanced diet. Turn the vegetables regularly to provide a range of nutrients.

b. Preparation: Vegetables ought to be finely slashed or destroyed to make them more straightforward for the bearded dragon to eat and process.

c. Supplementation: Grown-up bearded dragons require calcium and lack vitamin enhancements to forestall. Dust the vegetables with a calcium supplement a couple of times each week to guarantee that the bearded dragon gets satisfactory calcium.

3. Mixed Diet

A mixed diet combines both insect based and plant-based foods to provide a balanced and nutritious meal plan for bearded dragons. This diet is suitable for both young and grown-up bearded dragons and guarantees that they get a variety of nutrients.

Sample Mixed Diet:

a. Insects: Offer crickets, dubia insects, or superworms 2-3 times each week for protein and fundamental amino acids.

b. Vegetables: Give a variety of vegetables, for example, kale, collard greens, squash, and bell peppers, day to day. Guarantee that the vegetables are finely hacked or destroyed for simple consumption.

c. Fruits: Offer fruits like berries, melons, and apples as periodic treats. Fruits ought to be fed with some restraint because of their high sugar content.

Feeding Guidelines:

a. Balance: Expect to balance the proportion of insects to vegetables in the diet to meet the nutritional necessities of the bearded dragon. For example, you could feed insects 2-3 times each week and vegetables day-to-day.

b. Supplementation: Keep on enhancing the diet with calcium and lacks of vitamin enhancements to forestall and guarantee ideal health.

CHAPTER FIVE

SIGNS OF NUTRITIONAL IMBALANCE

Nutritional imbalances in bearded dragons can prompt different health issues. It's critical to perceive the signs ahead of schedule to give fitting care and change their diet as needed. Here are a few common indications of nutritional imbalance in bearded dragons:

Metabolic Bone Disease (MBD)

One of the most common problems coming about because of nutritional deficiencies is metabolic bone disease (MBD). This condition happens because of an absence of calcium, phosphorus, or vitamin D3, which are fundamental for bone health.

Signs of MBD include:

1. Soft or rubbery jaws and limbs
2. Swollen or disfigured limbs
3. Difficulty strolling or hauling limbs
4. Tremors or shaking
5. Fractures or breaks in the bones

Lack of vitamin A

Vitamin A is pivotal for a bearded dragon's vision, development, and invulnerable system. A lack of this vitamin can prompt:

1. Swollen eyes
2. Runny nose or mouth
3. Loss of appetite

4. Lethargy or shortcoming

5. Skin problems or shedding issues

Obesity

Overfeeding or giving an imbalanced diet can prompt obesity in bearded dragons. Indications of obesity include:

1. Excessive weight gain

2. Difficulty moving or climbing

3. Fatty deposits around the tail and limbs

4. Reduced movement levels

Digestive Issues

Deficient or unseemly food can bring about digestive problems:

1. Diarrhea or loose stools

2. Constipation

3. Regurgitation or heaving

4. Lack of appetite

5. Bloated appearance

Signs of Nutritional Imbalance in Hatchlings and Adolescents

Young bearded dragons are more susceptible to nutritional imbalances because of their quick development. Signs to look for include:

1. Stunted development or growth

2. Weakness or lethargy

3. Deformed limbs

4. Soft or rubbery jaw

5. Difficulty shedding

Preventing Nutritional Imbalance

To prevent nutritional imbalances in bearded dragons, a balanced and changed diet is fundamental. A healthy diet for a bearded dragon ought to include:

1. Insects: Crickets, mealworms, super worms, and dubia insects are great sources of protein. Guarantee they are gut-loaded (fed nutritious food before being proposed to the dragon) and tidied with calcium and vitamin D3 supplements.

2. Greens and Vegetables: Offer a variety of mixed greens and vegetables, for example, collard greens, mustard greens, dandelion greens, and squash.

These provide fundamental vitamins and minerals.

3. Fruits: Fruits like berries, melons, and apples can be offered periodically as treats but ought not be an essential piece of their diet because of their high sugar content.

4. Supplements: Residue insects with calcium and vitamin D3 supplements 3–4 times each week and a multivitamin supplement once every week to guarantee they are getting every one of the fundamental nutrients.

5. UVB Lighting: Giving proper UVB lighting is urgent for bearded dragons to metabolize calcium and keep up with

healthy bones. Guarantee they approach UVB light for 10–12 hours per day.

FEEDING YOUNG VERSUS GROWN-UP BEARDED DRAGONS

Feeding young and grown-up bearded dragons requires various methodologies because of their shifting nutritional necessities. This is a guide en route to feeding each age group properly:

Feeding Young Bearded Dragons:

Young bearded dragons (as long as a year and a half old) are developing quickly and require a diet that is rich in protein and calcium to help their development and improvement.

1. Protein:

a.   Offer youthful bearded dragons a diet that is high in protein.

b.   Feed them little bugs like crickets, mealworms, and little insects every day.

c.   Ensure the bugs are gut-loaded (fed a nutritious diet) before offering them to your dragon.

2.   Calcium and Vitamin D3:

a.   Calcium and vitamin D3 are critical for proper bone advancement in young bearded dragons.

b.   Dust the bugs with a calcium supplement (without phosphorus), something like 3–4 times each week.

c.   Ensure your dragon approaches UVB light to assist with vitamin D3 union.

3.   Vegetables and Greens:

a.   Offer a variety of mixed greens and vegetables to youthful bearded dragons.

b.   Chop the vegetables finely and offer them every day.

c.   Examples of suitable vegetables include collard greens, kale, dandelion greens, and butternut squash.

4.   Fruits:

a.   Limit how much fruit is given to young bearded dragons, as they are higher in sugar.

b. Offer limited quantities of fruits like berries, mango, and papaya once in a while as treats.

Feeding Grown-up Bearded Dragons:

Grown-up bearded dragons (more than a year and a half old) have slower development rates and different nutritional necessities compared with young dragons. Their diet ought to be balanced with lower protein content and more vegetables.

1. Protein:

a. Offer grown-up bearded dragons a diet that is lower in protein than that of young dragons.

b.   Feed them insects like crickets, mealworms, and cockroaches 2-3 times each week.

c.   Ensure the insects are gut-loaded before offering them to your dragon.

2.   Calcium and Vitamin D3:

a.   Continue to give calcium enhancements to grown-up bearded dragons, but lessen the frequency to 2-3 times each week.

b.   UVB light is as yet fundamental for grown-up dragons to assist with vitamin D3 blend.

3.   Vegetables and Greens:

a.   Vegetables ought to make up most of a grown-up bearded dragon's diet.

b.   Offer a variety of mixed greens and vegetables every day.

c.   Examples of suitable vegetables include collard greens, kale, dandelion greens, bell peppers, and squash.

4.   Fruits:

a.   Fruits ought to in any case, be given sparingly to grown-up bearded dragons because of their high sugar content.

b.   Offer modest quantities of fruits like berries, mango, and papaya sometimes as treats.

CHAPTER SIX

HYDRATION AND WATER NEEDS

Bearded dragons, similar to all reptiles, need water to remain healthy, but they don't hydrate the same way vertebrates do. This is the very thing that you really want to be familiar with hydration and water needs for your bearded dragon.

In the wild, bearded dragons get the vast majority of their water from the food they eat. They're omnivores, and that implies they eat both bugs and plant matter. Crickets, mealworms, and vegetables like kale and collard greens all contain some degree of dampness that helps hydrate your dragon.

In any case, it's as important as ever to give a new stock of water to your bearded dragon. Despite the fact that they don't drink a lot, having water available is significant, particularly for young dragons and females that are laying eggs.

Here are a few hints to ensure your bearded dragon stays hydrated:

1. Water Dish: Consistently have a shallow dish of new water available in the enclosure. Ensure the dish is shallow enough for your dragon to effortlessly access it without the risk of suffocating. Supplant the water from day to day to keep it clean.

2. Misting: A few bearded dragons appreciate being moistened with water. You can fog their enclosure a few times

per day to assist with expanding humidity and permit them to hydrate drops off surfaces. This can be especially beneficial for more youthful dragons.

3. Bathing: Giving your bearded dragon a bath is one more method for empowering hydration. Fill a shallow compartment with tepid water (not hot!) and let your dragon drench for about 10–15 minutes. Many bearded dragons will drink while they're in the bath.

4. Hydrating Foods: Offer new fruits and vegetables with high water content, similar to cucumber, zucchini, and melons, as a feature of their diet. This gives them hydration as well as changing their diet.

5. Watch for Indications of Drying out: It's crucial to screen your bearded dragon's hydration levels. Indications of parchedness include depressed eyes, crumpled skin, lethargy, and a diminished appetite. In the event that you notice any of these side effects, it's critical to find prompt ways to expand your dragon's water consumption and consult with a veterinarian if necessary.

OCCASIONAL DIET CHANGES

Bearded dragons are omnivorous reptiles, and that implies they eat a blend of plants and insects. Their diet can change with the seasons, just as it does for some creatures.

In the hotter months, bearded dragons will generally eat more bugs. This is because bugs are more dynamic and easier to find throughout the spring and summer. Crickets, mealworms, and little bugs are a portion of their number one treats. Bugs are an extraordinary wellspring of protein for bearded dragons, helping them develop and remain healthy.

During the cooler months, bearded dragons eat fewer bugs and more plants. This is because there are fewer bugs around, and the dragons' metabolism dials back a bit in light of the cooler temperatures. Salad greens like collard greens, kale, and mustard greens become a bigger piece of their diet. These greens provide fundamental vitamins and minerals, assisting with keeping the dragons healthy in any event, when they're not as dynamic.

Regardless of the time, giving a fluctuating and balanced diet to your bearded dragon is important. This implies offering a blend of bugs, mixed greens, and different vegetables like bell peppers, squash, and carrots. You can likewise offer periodic fruits like berries or mango as a unique treat.

It's likewise important to remember that youthful bearded dragons have unexpected dietary necessities in comparison to grown-ups. Youthful dragons need more protein to help their fast development, so their diet ought to incorporate a higher level of bugs compared with grown-up dragons.

In conclusion, consistently make a point to give new water to your bearded dragon. While they get a great deal of their hydration from their food, it's still important to have a shallow dish of clean water available consistently.

PICKING AND RAISING LIVE FEEDERS

Picking Live Feeders

1. Crickets: Crickets are a well known decision and are promptly available. Pick crickets that are the right size for your bearded dragon, typically no longer than the space between their eyes. Stomach load crickets with nutritious greens like kale or carrots before feeding them to your bearded dragon to boost their nutritional worth.

2. Dubia Insects: Dubia cockroaches are high in protein and low in fat. They are not difficult to process and are less inclined to cause impaction, compared with different feeders. Begin with more modest insects and increment the size as your bearded dragon develops.

3. Mealworms: Mealworms are another common feeder, but they have a harder exoskeleton, which can be challenging for more young bearded dragons to process. Taking care of mealworms with some restraint and not as a staple food is best.

4. Waxworms: Waxworms are high in fat and ought to be fed sparingly as a periodic treat because of their high-fat content.

5. Greens and Vegetables: notwithstanding live feeders, offering a variety of greens and vegetables to your bearded dragon is fundamental. Collard greens, mustard greens, and squash are great choices.

Raising Live Feeders

1. Housing: Keep your live feeders in a different holder to guarantee they are healthy and liberated from parasites. Utilize a well-ventilated holder with egg cartons or paper towel rolls for hiding spots.

2. Feeding: Feed your live feeders a nutritious diet to guarantee they are healthy for your bearded dragon. You can utilize commercial cricket or

cockroach chow, enhanced with new fruits and vegetables.

3. Gut-loading: Gut loading your feeders is important to give your bearded dragon fundamental nutrients. Feed them nutritious foods like mixed greens, carrots, and commercial gut-load diets for no less than 24 hours before feeding them to your bearded dragon.

4. Supplementation: Residue your live feeders with a calcium and vitamin D3 supplement before feeding them to your bearded dragon to guarantee they get fundamental nutrients.

5. Cleaning: Regularly spotless and keep up with the feeder compartment to

prevent the spread of bacteria and parasites. Eliminate any uneaten food and waste immediately.

CHAPTER SEVEN

IMPORTANCE OF GUT LOADING AND DUSTING

Gut-loading and dusting are two significant practices for guaranteeing the health and well-being of your bearded dragon. While they could sound specialized, they're basically about feeding your pet a nutritious diet and enhancing it to ensure they get every one of the vitamins and minerals they need.

Gut-loading is the most common way of feeding the bugs that you intend to provide your bearded dragon with a highly nutritious diet. This implies that the bugs will be loaded with vitamins and minerals by the time your dragon eats them. Proper gut-loading foods incorporate dull salad greens, carrots, and commercial gut-loading diets that you can find at pet stores. The thought is that when your bearded dragon eats these bugs, they're not simply getting a meal; they're getting a nutrient-pressed super food that helps keep them healthy.

Dusting, then again, includes daintily covering the bugs with a reptile-explicit vitamin and mineral enhancement before feeding them to your bearded dragon. This guarantees that regardless of whether the bugs haven't been completely gut-loaded, your dragon is still getting every one of the fundamental nutrients it needs. You can buy reptile calcium and vitamin supplements at all pet stores, and it's as simple as gently shaking the insects in the powder before feeding them to your dragon.

Presently, you may be asking why this quarrel about what your bearded dragon eats. Well, bearded dragons require a balanced diet to remain healthy, very much like some other creature or even us people. In the wild, they'd eat a variety of bugs, plants, and, surprisingly, a few little creatures. In captivity, it's our responsibility to recreate this different diet as intently as could really be expected.

A well-fed bearded dragon will be more dynamic, have a stronger resistance system, and generally be more joyful and healthier. On the other side, a diet lacking in fundamental nutrients can prompt health problems like metabolic bone disease, which can be truly troublesome and even perilous.

FEEDING BEHAVIOR AND FRQUENCY

Feeding Behavior

Bearded dragons are diurnal, and that implies they are generally dynamic during the day. Their feeding behavior is regularly artful, meaning they will eat at whatever point food is available. In the wild, they consume a variety of bugs, like crickets, mealworms, and cockroaches, as well as vegetation, like mixed greens and fruits.

In captivity, it's crucial to offer a balanced diet that emulates their natural feeding habits. Youthful bearded dragons (under a year old) are basically predatory and ought to be offered a larger number of bugs than plants. As they get older, their diet shifts more towards plant matter, but they actually require a blend of both.

Feeding Frequency

The feeding frequency of a bearded dragon changes based on its age:

1. Hatchlings (0–3 months old): They ought to be fed 2-3 times each day. Offer some properly estimated insects, as they can eat in about 10–15 minutes during each feeding. It's important to provide a

variety of insects to guarantee a balanced diet.

2. Juveniles (3–12 months old): Feed them on more than one occasion per day. At this stage, their diet ought to in any case, consist fundamentally of insects, but you can begin presenting more mixed greens and vegetables.

3. Adults (1 year and older): Feed grown-up bearded dragons one time per day. Their diet ought to consist of about 80% plant matter and 20% bugs. Offer a variety of vegetables and mixed greens, for example, collard greens, mustard greens, and squash, alongside periodic fruits like berries or melon.

General Feeding Tips

1.   Gut loading insects: Before feeding insects to your bearded dragon, 'gut load' them by feeding them nutrient-rich foods like carrots, mixed greens, or commercial stomach stacking diets. This guarantees that the bugs are nutritious for your dragon.

2.   Cleaning with calcium: Residue the bugs with a calcium supplement 2-3 times each week and a multivitamin once per week to guarantee your bearded dragon is getting every one of the fundamental nutrients.

3.   Fresh water: Consistently give new, clean water in a shallow dish to your bearded dragon. While they principally

get their hydration from their food, they might hydrate infrequently.

4. Observation: Consistently screen your bearded dragon while feeding to guarantee they are eating properly and to eliminate any uneaten food to keep up with neatness.

CHAPTER EIGHT

COMMON NUTRITIONAL DEFICIENCIES AND SOLUTIONS

1. Calcium Deficiency (Hypocalcemia):

a. Symptoms: Shortcoming, quakes, jerking, enlarged limbs or jaw, and trouble moving.

b. Solution: Guarantee your bearded dragon's diet incorporates calcium-rich foods like dim mixed greens (e.g., collard greens, mustard greens), calcium enhancements, and gut loaded insects (insects fed with high-calcium foods before being fed to the dragon). Give a calcium supplement with vitamin D3 to support calcium absorption.

2. Vitamin D3 Deficiency:

a. Symptoms: relaxation of the bones, enlarged limbs or jaw, trouble moving, and absence of appetite.

b. Solution: expose your bearded dragon to natural daylight or utilize an UVB light to assist them with incorporating vitamin D3. Also, vitamin D3 enhancements can be given, but be careful not to over-enhance.

3. Vitamin A Deficiency:

a. Symptoms: Enlarged eyes, respiratory contamination, absence of appetite, and harsh skin.

b. Solution: Offer foods high in vitamin A, like carrots, yams, and squash. Commercial vitamin enhancements can likewise be utilized, but they ought to be utilized sparingly.

4. Protein Deficiency:

a. Symptoms: Hindered development, shortcoming, and decreased movement levels.

b. Solution: Guarantee your bearded dragon's diet incorporates a variety of gut loaded insects like crickets, cockroaches, and mealworms. Also, periodic deals with things like pinkie mice or waxworms can be given.

5. Dehydration:

a. Symptoms: Indented eyes, creased skin, lethargy, and diminished appetite.

b.   Solution: Give new, clean water day-to-day and fog the enclosure to keep up with humidity. Offer water-rich foods like cucumbers and mixed greens.

6.   Fiber Lack:

a.   Symptoms: Constipation, decreased appetite, and bloating.

b.   Solution: Offer a variety of vegetables and fruits, for example, bell peppers, squash, and berries, to guarantee a balanced diet with sufficient fiber.

Prevention Tips:

1.   Continuously give a changed diet consisting of vegetables, fruits, and gut-loaded insects.

2. Use calcium and vitamin supplements depending on the situation, but stay away from over supplementation.

3. Guarantee proper hydration by giving new water and keeping up with fitting humidity levels in the enclosure.

4. Regularly screen your bearded dragon's health and behavior for any indications of nutritional inadequacies.

FEEDING CHALLENGES AND SOLUTIONS

1. Inadequate Variety of Foods:

a. Challenge: Bearded dragons require a different diet to get every one of the nutrients they need. Feeding them a similar food regularly can prompt nutritional deficiencies.

b. Solution: Offer a variety of vegetables, fruits, and insects to guarantee a balanced diet. Examples of suitable vegetables include collard greens, mustard greens, and bell peppers. Safe fruits incorporate berries, mango, and papaya. Bugs like crickets, dubia insects, and mealworms are great protein sources.

2. Refusal to Eat Vegetables:

a. Challenge: A few bearded dragons might be hesitant to eat vegetables, favoring just bugs.

b. Solution: To empower vegetable consumption, have a go at offering finely hacked or ground veggies blended in with the most loved bugs. You can likewise attempt hand-feeding or offering veggies on a shallow dish to make them seriously engaged.

3. Overfeeding or Underfeeding:

a. Challenge: It very well may be hard to decide the perfect proportion of food to offer.

b. Solution: A decent guideline is to offer bugs that are no bigger than the space between the bearded dragon's eyes, and vegetables ought to be about the size of the dragon's head. Screen their body condition and change how much food they need.

4. Poor Calcium and Vitamin D3 Intake:

a. Challenge: Bearded dragons need calcium and vitamin D3 for proper bone development and health. Without these nutrients, they can foster metabolic bone disease.

b. Solution: Residue bugs with a calcium powder that likewise contains vitamin D3 before feeding. Give them a UVB light in their enclosure to assist them with creating vitamin D3 naturally.

5. Dehydration:

a. Challenge: Bearded dragons may not hydrate, particularly assuming they are fed fundamentally dry foods.

b. Solution: Offer a shallow dish of new water every day. You can likewise fog their vegetables with water to increase dampness consumption. A few bearded dragons like to drink from a dropper or needle.

6. Temperature and Lighting Issues:

a. Challenge: Inaccurate temperatures can influence a bearded dragon's appetite and processing.

b. Solution: Guarantee the basking spot comes to 95–110°F (35–43°C) and the cooler side of the enclosure stays around 75–85°F (24–29°C). Utilize a quality UVB light to give the essential range to processing and vitamin D3 combination.

CHAPTER NINE
SIGNS OF OVERFEEDING AND UNDERFEEDING

Overfeeding:

1. Obesity: One of the most visible signs is an overweight or obese appearance. A healthy bearded dragon ought to have a smoothed out body shape without extreme fat stores.

2. Fat Pads: Fat pads can be found on the sides of the neck, base of the tail, or limbs. These are delicate and soft to the touch.

3. Lethargy: Overfed bearded dragons might be less dynamic than expected. They could invest more energy in

basking and spend less time investigating or hunting.

4. Diarrhea: An expansion in the recurrence or consistency of their droppings can be an indication of overfeeding, as their digestive system battles to handle the overabundance of food.

5. Difficulty Breathing: In extreme cases, obesity can prompt respiratory problems because of the unnecessary fat pushing on their lungs.

6. Fatty Liver Disease: Overfeeding can prompt greasy liver disease, which can be hazardous. Side effects include lethargy, loss of appetite, and an enlarged abdomen.

Underfeeding:

1. Weight Misfortune: A noticeable reduction in body weight or a depressed appearance of the fat cushions can show underfeeding.

2. Reduced Movement: An underfed bearded dragon might become dormant and less dynamic. They probably won't move around so much and may invest more energy in hiding.

3. Weakness: They could experience issues moving or climbing, and their muscles might show up less evolved.

4. Brittle Bones: An absence of proper nutrition can prompt metabolic bone disease, making the bones powerless

and brittle. This should be visible in bowed or disfigured limbs.

5. Dull Skin and Shedding Issues: Underfed bearded dragons might have dull, dry skin and may battle with inadequate or problematic sheds.

6. Decreased Appetite: In the event that a bearded dragon is underfed, they could eat less or show no interest in food by any means.

Balanced Feeding:

To guarantee your bearded dragon is getting the perfect proportion of food, offer them fittingly estimated prey things (like crickets or cockroaches) that are no bigger than the space between their eyes. Adolescents ought to be fed all the more habitually, while adults can be fed on rare occasions. New vegetables ought to likewise be offered every day, and calcium and vitamin enhancements ought to be utilized as coordinated.

Regularly checking your bearded dragon's weight, movement level, and appetite can assist you with changing their diet depending on the situation to keep them healthy and cheerful. Continuously consult with a veterinarian in the event that you're uncertain about your bearded dragon's feeding or, on the other hand, assuming you notice any unsettling side effects.

THE END

www.ingramcontent.com/pod-product-compliance
Lightning Source LLC
Chambersburg PA
CBHW070315230526
45470CB00002B/888